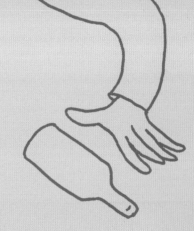

The EVERYDAY JOURNEYS of ORDINARY THINGS

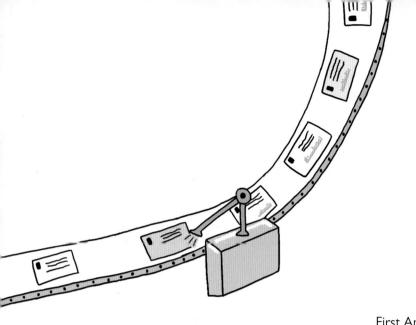

First American Edition 2019
Kane Miller, A Division of EDC Publishing

Copyright © 2019 Quarto Publishing plc

Published by arrangement with Ivy Kids, an imprint of The Quarto Group.
All rights reserved. No part of this book may be reproduced, transmitted
or stored in an information retrieval system in any form or by any means, graphic,
electronic or mechanical, including photocopying, taping and recording,
without prior written permission from the publisher.

For information contact:
Kane Miller, A Division of EDC Publishing
PO Box 470663
Tulsa, OK 74147-0663
www.kanemiller.com
www.edcpub.com
www.usbornebooksandmore.com

Library of Congress Control Number: 2018942374

Printed in China

ISBN: 978-1-61067-729-5

1 2 3 4 5 6 7 8 9 10

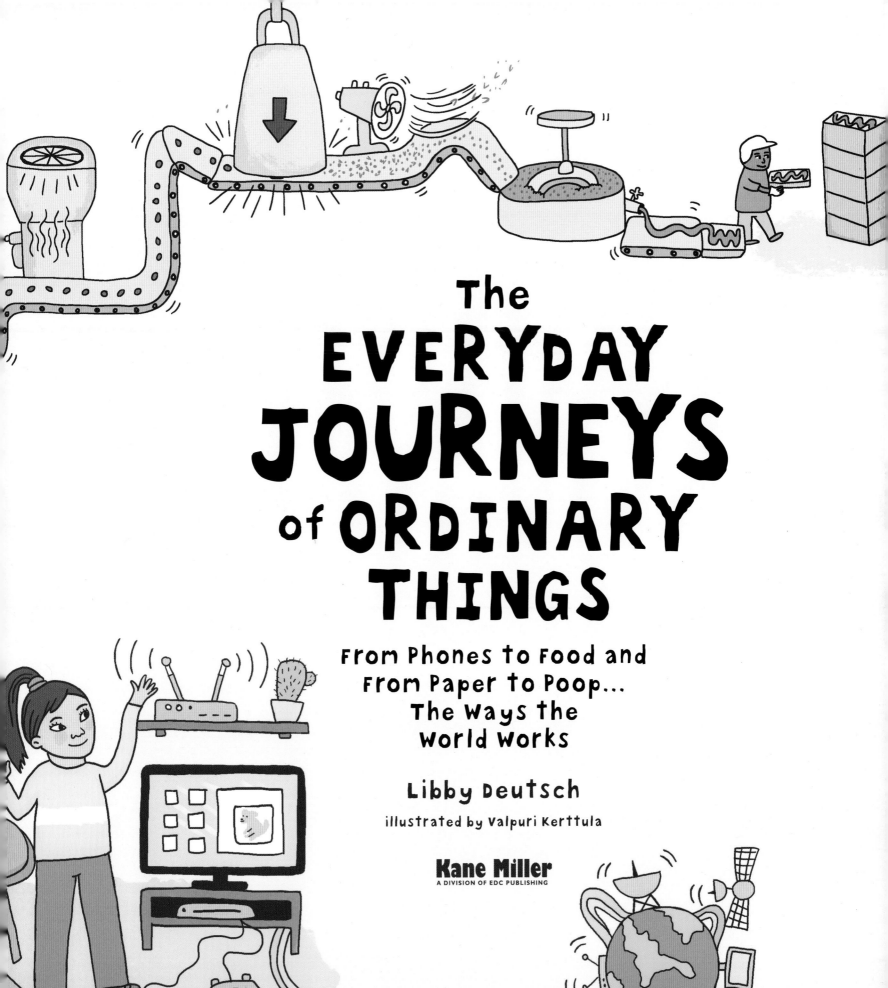

The EVERYDAY JOURNEYS of ORDINARY THINGS

From Phones to Food and From Paper to Poop... The Ways the World Works

Libby Deutsch

illustrated by Valpuri Kerttula

Kane Miller
A DIVISION OF EDC PUBLISHING

CONTENTS

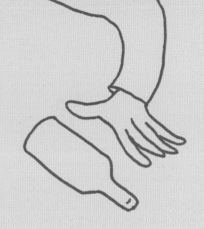

Introduction 6

8 **How Does Mail Reach Me?**
The journey made by a birthday card

Does Food Grow on Shelves? 10
The journey of a banana

12 **Where on Earth Are You?**
The journey of GPS

How Are My Jeans Made? 14
The journey of jeans

16 **From Tree to Paper**
The journey of paper

From Idea to Bookshelf 18
The journey of a book

20 **At the Click of a Button**
The journey of your online shopping order

How Glass 22
Can be Recycled
Over and Over Again
The journey of a used glass bottle

24 **From Tree to Shelf**
The journey of chocolate

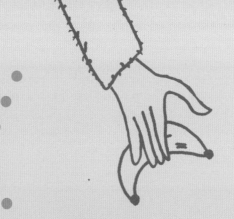

26 Where Does My Suitcase Go When I Fly On a Plane?
The journey of a suitcase

Where Does the Water in the Faucet Come From? 28
A journey through the water cycle

30 From Fossil to Car
The journey of gasoline

The Invisible Movement of Millions 32
The journey of money

34 Lights, Camera, Action!
The journey of a movie from script to screen

How Does the Internet Work? 36
The superfast journey of information

38 From Studio to Stereo
The journey of music

Where Do the Words Go? 40
The journey of a phone call

42 How Does Electricity Reach My House?
The journey of electricity

What Happens When I Flush? 44
The journey of poop

46 From Cow to Carton
The journey of milk

Fifteen minutes of ordinary things. But just to make them happen, somewhere in the world, millions-of-years-old sludge has been pulled from the ground and burned; cotton seeds have been planted; wood pulp has been transformed into paper by machines bigger than a three-story building; digital information has soared into space and bounced back again; poop has been sieved, and billions and billions' worth of invisible money has whizzed around the world...

Every day we are surrounded by ordinary objects and services that have been on the most extraordinary journeys to reach us; journeys that are all happening RIGHT NOW. Take a look around you. What do you see? Where does it come from? How did it begin?

Once you start noticing these extraordinary journeys, you'll never look at the world in the same way again...

A word about the machines

In this book, there are lots of cool machines. The pictures of the machines are drawn as accurately as possible, but sometimes, the pictures are a representation of what the machine really looks like— like the magnet on the right.

How Does Mail Reach Me?

The journey made by a birthday card

If a friend mails you a birthday card, a whole team of people using vans, trucks and planes will make sure it reaches you. Here's how it works...

1 First, your friend writes your address on the envelope and puts the card in a mailbox.

2 A mail carrier empties the mailbox and takes the letters to the nearest post office.

3 A truck collects the mail from the post office and takes it to a mail processing plant.

4 The letters are tipped onto a conveyor belt. Any larger items, such as packages, are filtered out and sorted separately. This is known as "culling."

5 Mail is fed into a special sorting machine that uses a camera to read handwritten addresses. It also flips the envelopes right-side up.

6 Another machine sprays each letter with a fluorescent bar code. The bar code contains information about your letter, including its zip code and delivery address.

The sorting machine can process 30,000 pieces of mail in one hour.

7
The mail is loaded onto a truck and taken to a processing plant closer to your neighborhood. Depending on where you live, it could be driven or flown there.

8
At the next processing plant, letters are fed into a machine that reads the bar codes. It sorts letters into a delivery order for the mail carrier at a local post office.

9
From the local post office, batches of letters are handed over to mail carriers, who deliver them right to people's front doors, even if they live somewhere very remote.

Every day, millions and millions of items of mail are delivered in this way.

Postal services around the world use almost every vehicle you can think of to deliver letters.

10
The mail carrier has to reach your house... ...no matter where you live.

Does Food Grow on Shelves?
The journey of a banana

Grocery stores are full of all kinds of food. But where does it all come from? The answer is: all over the world. Let's take a look at the journey made by one of our favorite foods—bananas.

1 Bananas come mostly from countries around the equator. These countries have a tropical climate where it's warm year round, with lots of rain.

Each large cluster of bananas is called a "banana stem." One stem is made up of several tiers (or levels) of bunches of bananas. The correct name for a bunch of bananas is a "hand," and a single banana is known as a "finger."

2 Bananas grow on a crop farm called a plantation. They grow in big, heavy bunches. Farmers hang plastic bags around the bananas while they are growing to protect them from wasps and birds.

3 Bananas are harvested when they have grown to a certain size but are still green. Most people don't like buying bruised bananas, so the fruit must be handled with great care. Foam pads are placed between the bunches to protect them.

4 The workers carry the stems from the tree and hang them on a trolley rail. When all the stems have been harvested, workers pull them along the rail to the sorting depot.

10

Where on Earth Are You?
The journey of GPS

Wherever you are, satellite navigation systems can locate you. But how does that information reach you when you might be up a mountain, trekking through a desert or sailing on the vast ocean?

1

Yes! You're going to see your favorite band, Which Direction?, with your best friend. Your grandma is taking you both, but she doesn't know exactly where the concert is—or even exactly where you are! She takes out her smartphone and goes to her navigation app.

2

Meanwhile, approximately 27 satellites are orbiting Earth, each equipped with an onboard computer, radio transmitter and very precise atomic clock. They continuously transmit information about their exact location and time via radio waves. These waves are traveling toward Earth at the speed of light. Together, the satellites form the Global Positioning System, or GPS.

Each satellite follows an exact route—like a bus—and travels around Earth twice a day. Their routes and timings have been set so that there are always at least three satellites within range of your phone receiver. They also transmit a list of where they should be and when—like a bus schedule!

GPS satellites are not the only objects orbiting Earth. There are hundreds of thousands of pieces of space rubbish, including a wrench and an astronaut's glove!

8

You have arrived. Your grandma switches off the navigation app on her phone. When the concert is over, she can use it to find the way home again. Have fun!

7

Suddenly, the navigation app on your grandma's phone starts to work! The control station's broadcasts have reached it. Her receiver is also picking up information from a fourth satellite, so it gives a more accurate reading.

3

As soon as your grandma turns on her navigation app, it begins to receive the information sent by the satellites.

• Each satellite transmits its precise location and the exact time to the nearest nanosecond.
• The receiver on the phone subtracts the time that the information was transmitted from the time it was received.
• Knowing that the information traveled at the speed of light, your receiver can tell your distance from the satellite.

The calculation:
Transmission time x speed of light = distance from satellite

4

Knowing how far you are from one satellite doesn't tell you where you are. But your grandma's receiver is getting this information from three different satellites. A point in each of the three readings will be the same—that's where your grandma's phone is.

5

Hmm, something's not working on your grandma's app. It keeps telling her she is in the wrong place. This is because some of the signals the phone is picking up have been delayed. First, they were slowed down by the electromagnetic field surrounding Earth, and also the bad weather. Then, because they can't travel through solid objects, like the tall buildings around you, they are taking a while to get through. The app doesn't know this though. It makes its calculations assuming that the signal traveled to you at the speed of light.

6

A local control station has picked up that the transmissions from one satellite are getting delayed. The control station is in a fixed position and has the list that says where each satellite should be and when, so it can tell when a transmission from a satellite is wrong. The control station recalculates the satellite's information and rebroadcasts it to the surrounding area.

The satellites are solar powered, and usually last for about ten years. When they are updated, the new model will have new features and the latest technology on board.

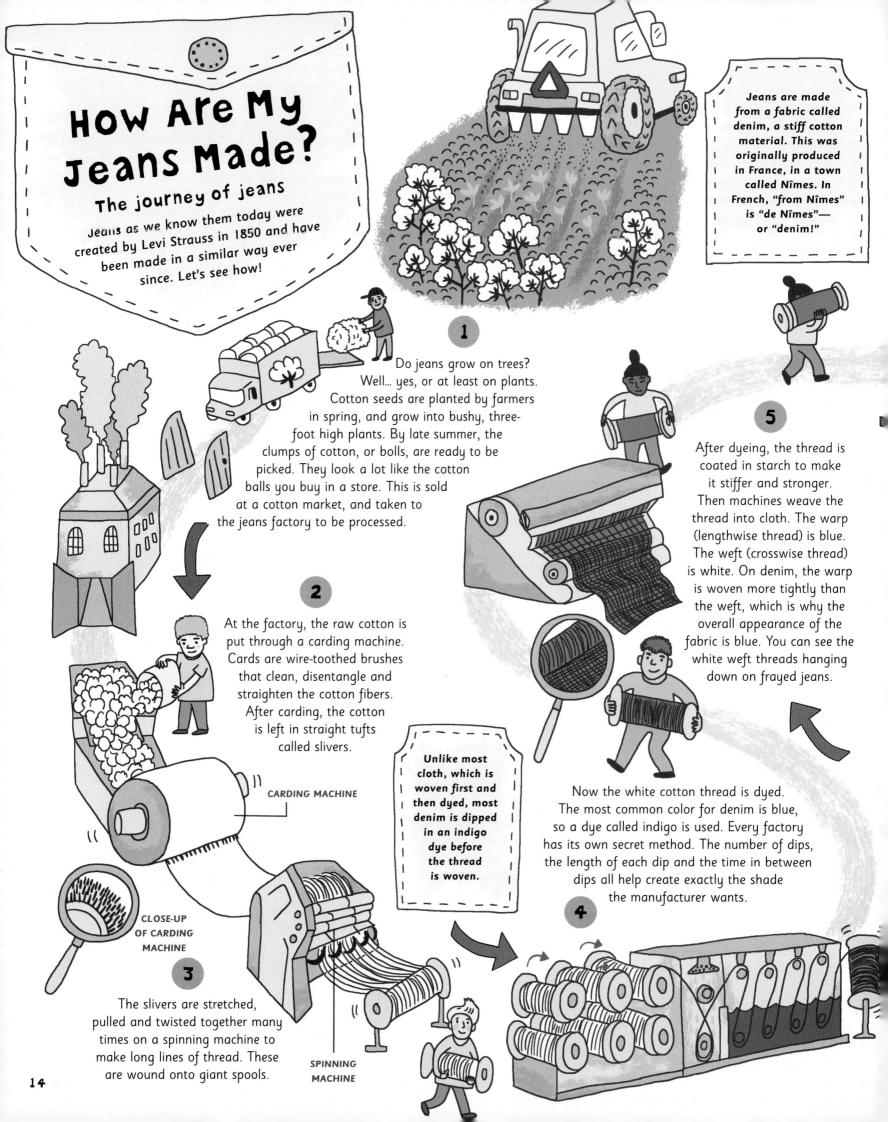

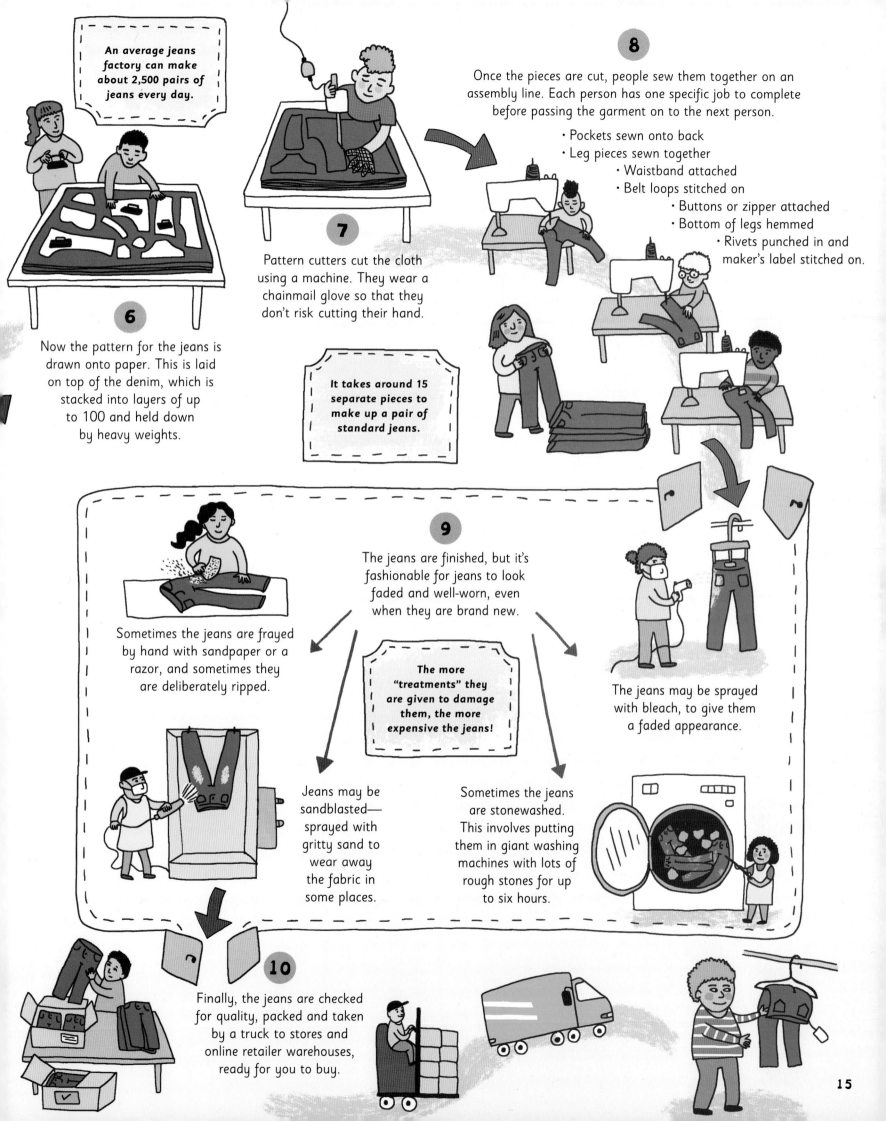

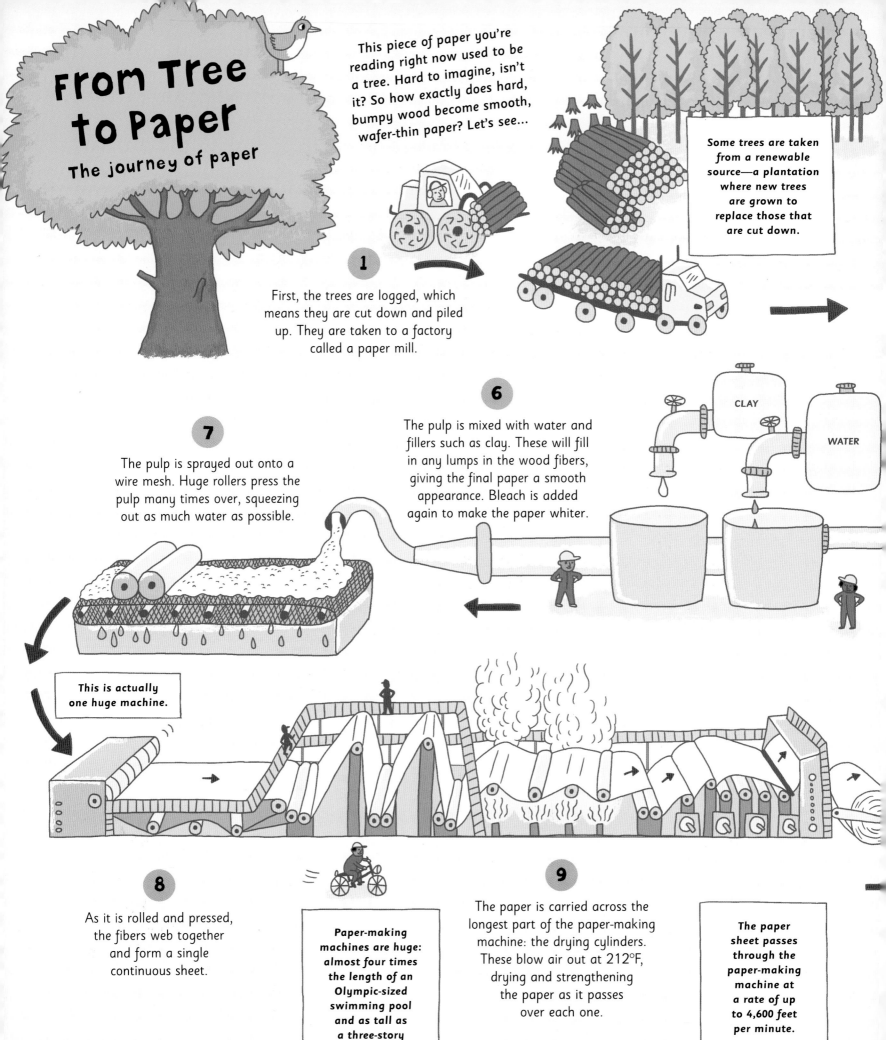

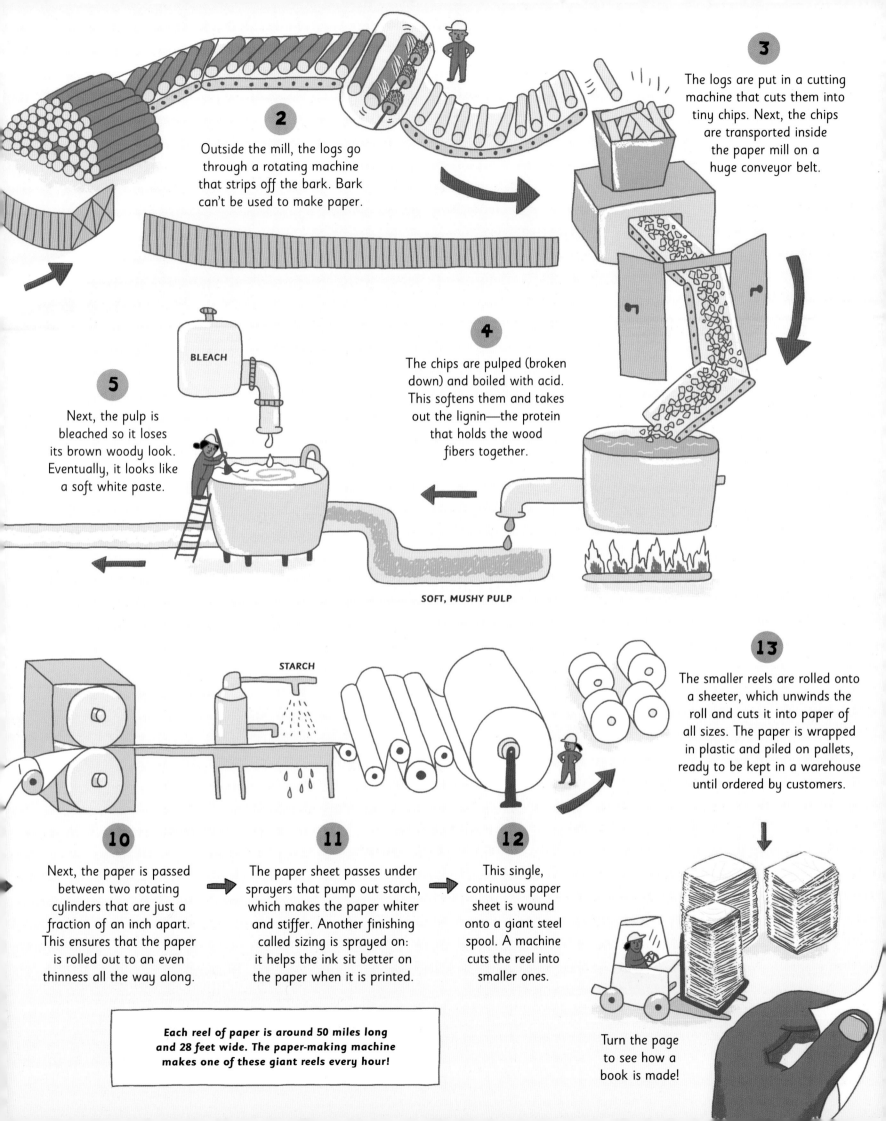

From Idea to Bookshelf

The journey of a book

What's your favorite book? Has it ever struck you that you are reading the very words the author wrote down, maybe late at night, sitting in their office? It's almost as though the story jumps straight from their head to yours. But what journey does the tale take in between?

1 First, the author has a great idea for a story. She writes it all down. Then she reads it back and rewrites it, cutting parts she doesn't like and adding new sections. This stage can take up to a year or longer.

At last— a GREAT book. I want to represent this writer!

2 The story has been written. Now the author wants people to read it! Some authors publish books themselves, but it costs a lot of money, so most try to find a publisher. First, they send it to an agent. The agent understands the publishing industry, finds good authors for publishers, and makes sure that authors get paid properly.

3  The agent sends the book to the right people at different publishing houses. This is a picture book, so she sends it to children's book editors.

4 At the publishing house, an editor reads the book. The editor knows a good story and can tell if other people will want to read it. He loves the book too!

5 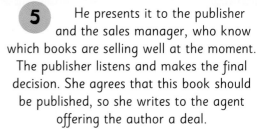 He presents it to the publisher and the sales manager, who know which books are selling well at the moment. The publisher listens and makes the final decision. She agrees that this book should be published, so she writes to the agent offering the author a deal.

18

6 Once the author, agent and publisher agree on a deal, they sign a contract that gives the publishing house the right to publish and sell the book.

The author is given a payment for the book, called an advance, and a percentage of every book sold, called a royalty.

7 The editor works on the book with the author to make the story as good as possible. This may include cutting parts or adding new bits.

14 As soon as the books arrive in the warehouse, they are sent out to bookstores and libraries, where the staff unpack the books and put them on their shelves, ready for you to enjoy!

9 The cover designer works on making the cover look great. People really *do* judge a book by its cover, so a lot of time is spent getting it right. Then, the editor writes the back cover text. The cover has to be seen by everyone—especially the sales team—who knows what customers want.

8 An illustrator selected by the publisher draws the pictures. Then, a designer puts the pictures and text in the right places on the pages. The author and illustrator often don't meet, even when they are working on a book full of illustrations.

13 If the printer is in a different country, the finished books will be sent on a ship across the ocean. This can take a number of weeks. Finally, the books arrive at a port. They are unloaded and taken in a truck to a big warehouse.

10 The last person to check the text is the proofreader. He looks at every single detail of the book to make sure it is right: spellings, page numbers, paragraph indentations—everything!

11 The production controller sends the computer files to the printer with instructions about how the book should be printed. The printer makes proof, or test, copies for the publisher to make sure that everything is correct. Then, the production controller gives the printer the go-ahead to print all of the books.

12 The pages of the book are printed on giant rolls of paper, which are cut by special machines. The books are bound together with glue and thread.

At the click of a Button

The journey of your online shopping order

When you buy something in a store, you take it to the cashier, pay for it and take your purchase home. But how does it work when you buy online? Let's look at what happens when you buy a birthday present for a friend.

4 Now the OMS communicates again through the server to your web browser to let you know that you have successfully purchased your item.

3 If your item is available, the OMS communicates with the server of your dad's bank, asking for the money. The bank responds, telling it that money is available.

STORE'S SERVER

BANK'S SERVER

1 You've found the perfect present online—a basketball. You click the "Buy" button and add your delivery address. Your dad enters his credit card details to make the payment.

2 Your web browser communicates with a server (a very powerful computer) that manages the store's website and Order Management System, or OMS—the store's main computer system. The OMS checks to see whether the present you want to buy is available—most likely somewhere miles away in a large warehouse. If it is not available, an order for more is automatically sent to the supplier.

The OMS will use your zip code to look for your item in a warehouse as close to your home as possible.

13 In a day or so, your doorbell rings. A delivery driver is standing on your doorstep with your present, ready to wrap and take to your friend's birthday party!

12 The parcels are packed onto trucks and taken to a shipping company. An email is sent to let you know your order has been dispatched.

20

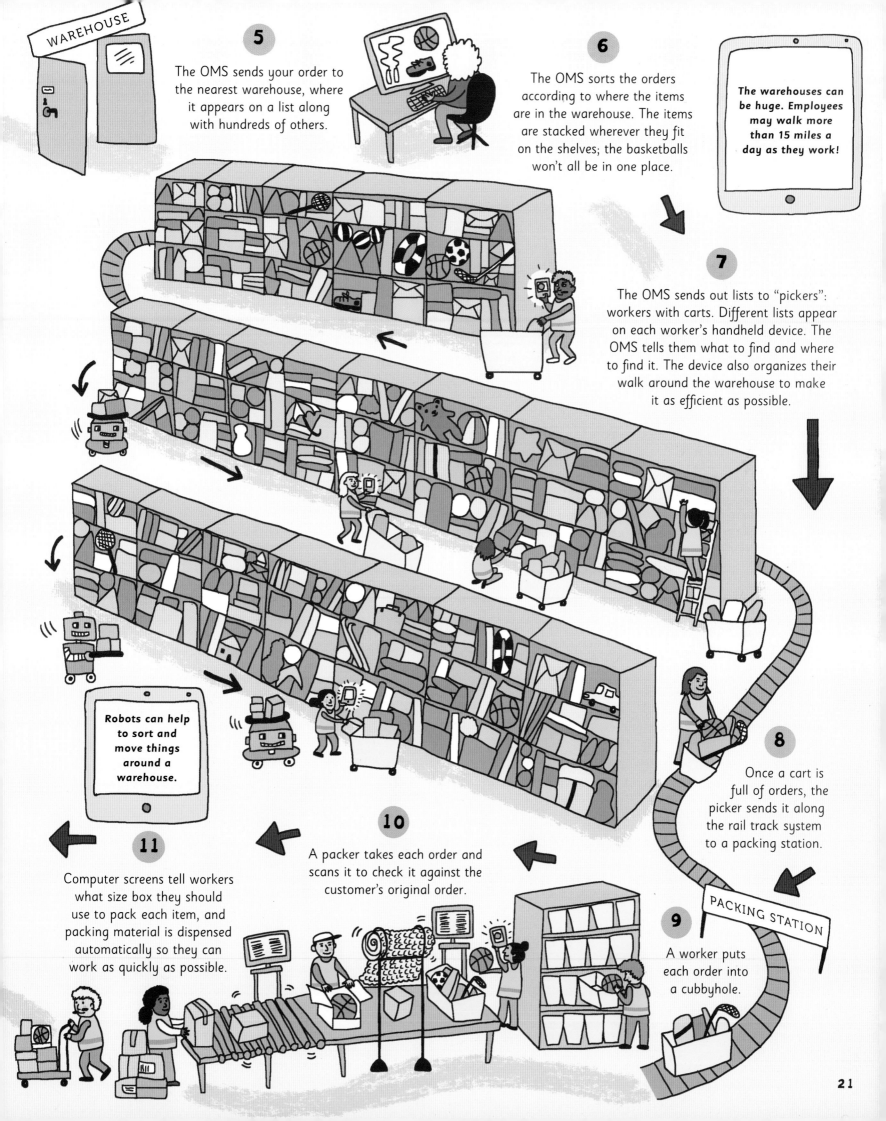

How Glass can be Recycled Over and Over Again

The journey of a used glass bottle

A glass bottle you recycle can be made into new bottles forever. Turning it back into new glass saves resources, energy and money. Here's what happens.

1 The garbage collectors pick up your recycling and take it to a materials recovery facility (MRF), where the glass, metal, paper and plastics are separated.

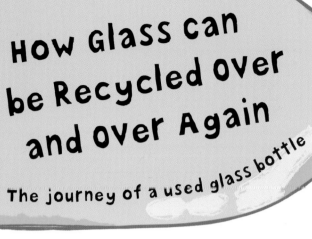

The energy saved from recycling just one bottle can power a computer for 25 minutes!

13 After they cool, the bottles are sold to a lemonade factory, where they are filled, lidded and sent to the store for you to buy, use and recycle all over again.

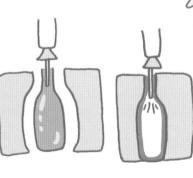

11 At a glass bottle factory, the cullet is melted in a furnace at temperatures of around 2,700°F (about six times hotter than a hot oven at home). Materials, including sand, are added to turn the glass into liquid.

12 The liquid glass is divided into soft cylinder shapes called gobs, which are dropped into molds. Air is blown into the middle of the gob, hollowing out the inside and forming the shape of the bottle.

Where Does My Suitcase Go When I Fly on a Plane?
The journey of a suitcase

You've got your tickets, packed your bags and you're at the airport, ready for your vacation—how exciting! Your suitcase will be taking a little trip of its own...

1

When you arrive at the airport, you will have baggage—a large bag containing items to take on your trip. The baggage needs to go in the hold of the plane—a big space in the belly of the plane. You may also have a carry-on with a few things you'll need during the flight.

Many airports have self-service check-in points too.

2

The check-in agent registers that you have arrived for your flight. They also weigh your bag. You will have been told how heavy your bag is allowed to be when you bought the plane tickets. If all the passengers turned up with very heavy bags, the plane wouldn't be able to get off the ground!

3

The agent attaches a tag with a barcode to your bag. This has all the information about your flight on it, so the baggage handlers know which plane to put it on.

8

Next, each bag travels through a large X-ray machine to be screened and checked for dangerous objects.

9

Airports are enormous places, and sometimes baggage has to go a long way to get from the check-in area to its plane. If so, it might be tipped into a little cart and whisked along at high speed on tracks that go up and down like a mini roller coaster. Wheee!

At this point, "transfer luggage" is added into the system. This is the baggage belonging to passengers who are changing planes at this airport to join your flight.

10

The cart tips your bag from a high conveyor belt down a slope, a bit like a curly slide, to the "reconciliation" area at the bottom. There, an employee scans the barcode to match your bag to your flight, and loads it onto a large container.

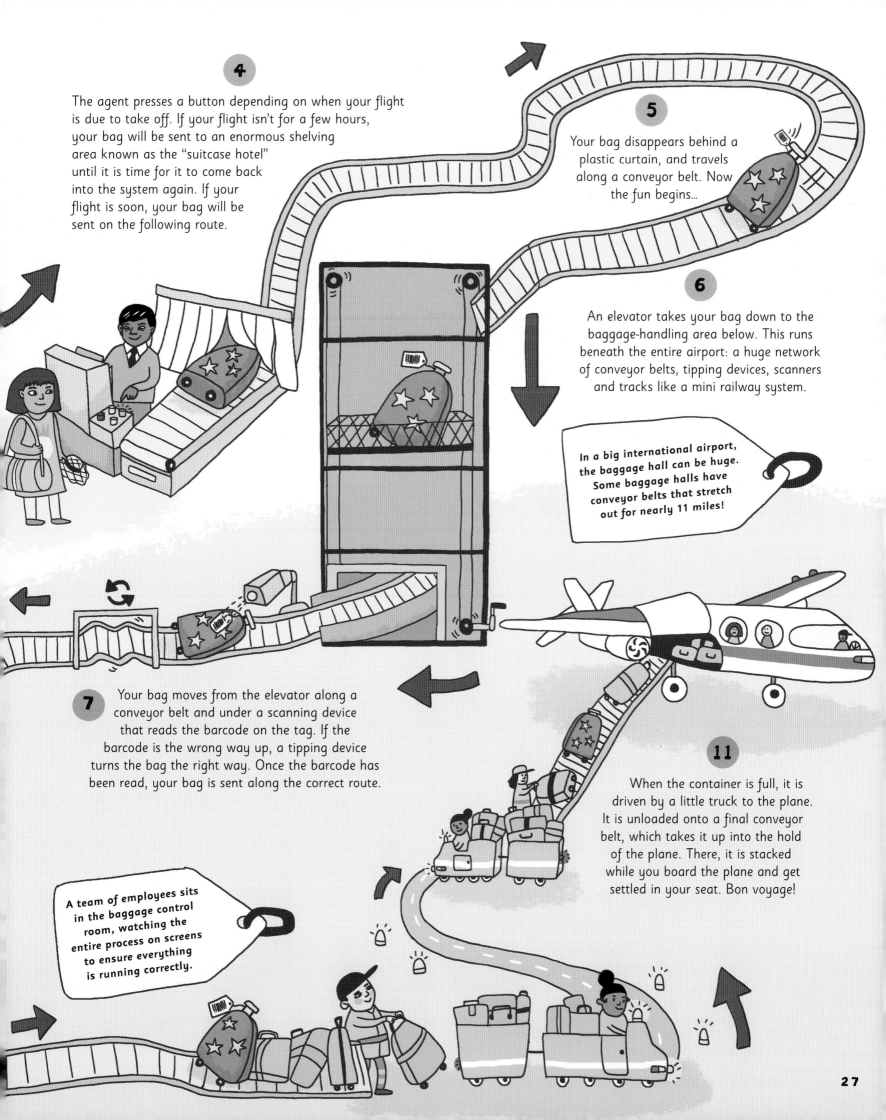

4 The agent presses a button depending on when your flight is due to take off. If your flight isn't for a few hours, your bag will be sent to an enormous shelving area known as the "suitcase hotel" until it is time for it to come back into the system again. If your flight is soon, your bag will be sent on the following route.

5 Your bag disappears behind a plastic curtain, and travels along a conveyor belt. Now the fun begins…

6 An elevator takes your bag down to the baggage-handling area below. This runs beneath the entire airport: a huge network of conveyor belts, tipping devices, scanners and tracks like a mini railway system.

In a big international airport, the baggage hall can be huge. Some baggage halls have conveyor belts that stretch out for nearly 11 miles!

7 Your bag moves from the elevator along a conveyor belt and under a scanning device that reads the barcode on the tag. If the barcode is the wrong way up, a tipping device turns the bag the right way. Once the barcode has been read, your bag is sent along the correct route.

11 When the container is full, it is driven by a little truck to the plane. It is unloaded onto a final conveyor belt, which takes it up into the hold of the plane. There, it is stacked while you board the plane and get settled in your seat. Bon voyage!

A team of employees sits in the baggage control room, watching the entire process on screens to ensure everything is running correctly.

From Fossil to car
The journey of gasoline

Have you ever heard of fossil fuels? A fuel is a material that is burned to produce heat and power. A fossil is the remains of a prehistoric plant or animal that has become embedded in layers of rock. We use a substance that is millions and millions of years old to run some of our most modern technology.

1 Between 550 and 65 million years ago, plants and animals died and sank to the bottom of the ocean. Over time, they were covered by mud.

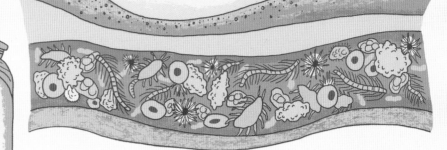

2 The mud preserved them for many thousands of years. In that time, oceans dried up, volcanoes erupted, and the mud became a layer of rock. The preserved organic (plant and animal) matter was trapped. It decomposed, slowly, until millions of years later it had become a sticky black liquid called crude oil. The oil produced a gas called natural gas. Both the oil and gas smell a bit like rotten eggs.

Crude oil and natural gas provide most of the fuel the world runs on today. But first we need to get to them—and sometimes they can be as far as 8 miles under the ground!

3 Crude oil is worth a lot of money. It is found around the world, and oil companies spend a lot of time and effort looking for it. Geologists use "seismic" technology: shock waves sent into the ground that bounce off the layers of rock and back to the surface, indicating what lies below.

4 Once companies have found oil, they build large platforms called oil rigs. These may be on land or at sea. If there is a huge supply, oil could be pumped there for many years. An enormous drill is used to penetrate deep into the ground.

5 The drill uses a diamond tip, one of the hardest substances on Earth, to penetrate through the rock layers. As it whizzes around, the friction creates a vast amount of heat. This would be dangerous if it hit explosive gas, so a kind of muddy water is continually pumped down through the rock to cool everything down.

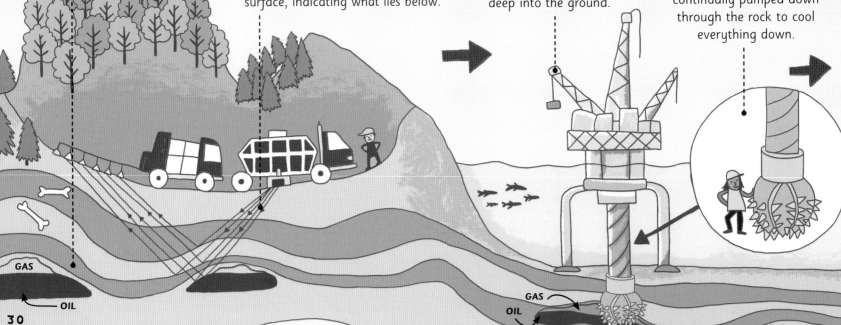

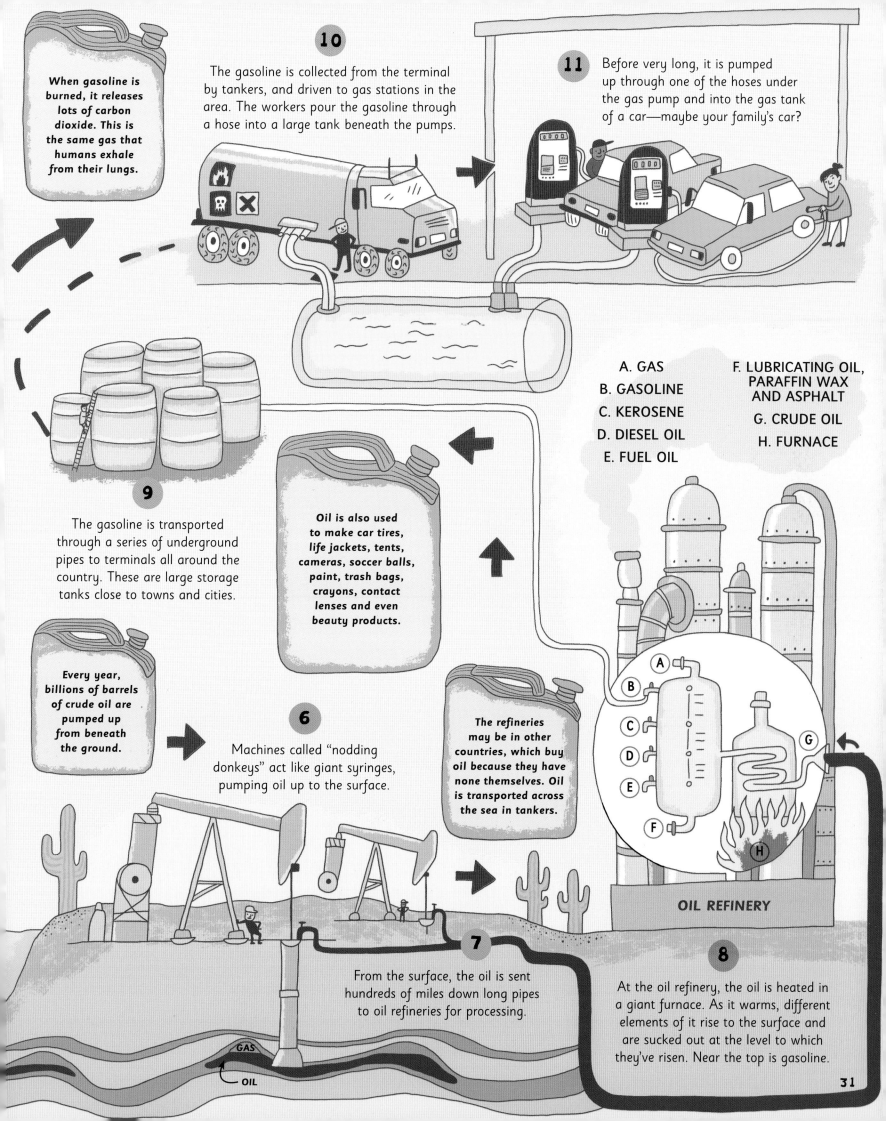

The Invisible Movement of Millions
The journey of money

Would you believe that money in today's online world is just an idea that we all believe in? Let's see how we got here.

1 Money is a currency, which is something you can use to trade with. Thousands of years ago, we traded with commodities—useful things. People might have traded a bag of salt for some beans or a cow for a tool. This was called commodity currency.

2 But cows died and beans could go moldy. Over time, swapping things was replaced by swapping gold. Gold was a metal available all over the world, it didn't go bad and was soft enough to form into coins. Silver soon joined it. This was known as coin currency.

3 But metal was heavy to carry around, so people left it with gold traders, who gave them a receipt for it. If your receipt was for five silver coins, you could trade it for something worth that much. This was called paper currency.

4 Gold and silver traders grew to become banks that looked after people's money for them. The customer could withdraw it when they needed, or deposit more. These transactions (movements of money) were written in account books so everyone could keep track.

> The British pound originally meant a pound (lb.) in weight of silver. So a £5 note represented 5 pounds of silver.

5

These days, account books have been replaced by sophisticated computer systems. You don't need to see the gold or silver those numbers represent any longer; you don't even need to see the paper money. The currency no longer needs to be visible or even exist as a physical object. You can transfer money electronically. Let's look at an example of how money is transferred electronically.

> Stores, employers, skilled workers and even governments use electronic bank transfers to trade billions of dollars every day. It's estimated that only 8 percent of the world's currency exists as physical cash. Most money is just an idea! Imagine what might happen if people stopped believing in it.

6 MONDAY

Your mom goes to work every day. At the end of the month, her employer's bank tells your mom's bank that it has paid her. The balance (amount of money) goes up in your mom's bank account and down in her employer's account, but no bills or coins have been exchanged. This is an electronic bank transfer.

ITEM	DEBIT	CREDIT	BALANCE
Work		$2,000	$2,000
Electricity	$50		$1,950
ATM withdrawal	$20		$1,930
Store	$30		$1,900
Shoe store	$25		$1,875

7 TUESDAY

Your mom needs to pay some bills— say, the electricity. The electricity company informs the bank how much your mom should pay and the bank does another electronic bank transfer. The balance goes down in her bank account and up in the account of the electricity company.

10 FRIDAY

Your mom goes online to buy you a new pair of school shoes. She enters the details of her card on the website page, which tells her bank that she wants to transfer some of her money to the online store. Since being paid, she has spent more than $100, and most of the money has been exchanged electronically.

9 THURSDAY

You and your mom go shopping, and she uses her card. She spends $30. The store's computer system communicates with the bank's computer system, and the money is transferred from her bank account to the store's account.

8 WEDNESDAY

Now your mom needs to pay $20 for your sports club. The club asks for cash. She goes to an Automated Teller Machine (ATM) with her card. When she puts in the card, the ATM reads the magnetic strip. Once she enters the correct Personal Identification Number (PIN) code, she accesses the computer system of her bank. The money taken out at the ATM is recorded by her bank.

Lights, camera, Action!

The journey of a movie from script to screen

Behind almost every movie you see are months—even years—of hard work by a team of people. Follow the journey of a movie idea.

1 This writer has a great idea for a movie. But he can't make it all by himself. He needs to pay for actors, costumes, set design, camera and lighting equipment, travel...

2 The writer has a meeting with a producer, and explains all about his idea for the movie. This is called a "pitch."

7 The movie is advertised, and the stars give interviews to promote it. There may be a premiere—a first showing of the movie with a party. The writer's idea has traveled from his imagination to the imagination of other people, and has become a reality!

6 Now all the footage has to be put together in the right order, using the best bits, taking out the parts that didn't work so well, and adding special effects, music, credits and any voice-over recordings. The film editor and the director do this work on computers.

A movie may be released in different parts of the world at different times, in theaters or straight to DVD or on-demand providers.

Cast List

WRITER | PRODUCER | DIRECTOR | DIRECTOR OF PHOTOGRAPHY (DP) | ACTOR 1 | ACTOR 2

3 If the producer likes the idea, they agree to make a sample of the movie. This is called a "treatment." Movies cost a lot of money to make so the producer needs to persuade people to invest money in it.

4 The producer has found the money, has the screenplay (the script with instructions for the actors) from the writer, but now needs to hire a director, a casting director and actors.

5 Finally it's time to shoot the movie! The producer hires a huge crew of people. They all have a different job to do on set. This includes camera operators, makeup artists, costume designers and sound and lighting engineers. Shooting the movie might take a whole year!

The scenes of the movie are not filmed in the order in which they appear. If there are several scenes set in the same place, these will be filmed at the same time, even if they are from different parts of the story.

How Does the Internet Work?

The superfast journey of information

The Internet is the network that links all of the computers in the world to each other. It's made up of all kinds of computers, tablets and smartphones. Say you want to find a picture on the Internet of your favorite dog, the labradoodle. How does the photo reach you?

1 In a search engine, such as Google, you type in "labradoodle" and click on "images." You see a cute picture that you want to download and click on it.

3 Your request for the labradoodle picture is sent to your Internet Service Provider (ISP), the company you pay for your Internet service. The ISP has lots of servers to store and transfer data. Servers are very powerful computers, which contain far more information than our home computers. They can be anywhere in the world.

The Internet is the name for the network of all the computers in the world. The World Wide Web is all the information on those computers that travels around the Internet.

2 Your computer, phone or tablet is linked to the Internet through a modem, a jumping-off point to the worldwide network of computers.

Every computer and server has an address of ten numbers called an Internet Protocol (IP) address.

4

Your ISP sends a request to the IP address of the server that has the image. It might be in another country. It finds the speediest way to reach the server with that IP address. Your request might travel along fiber-optic cables under the ocean or via a satellite in space. Along the way, your request goes through more computers called routers, switches or hubs, which are like stations on the Internet.

You can look up an IP address on a Domain Name Server, which is like a giant address book.

5

All information on the Internet is broken down and sent in small "packets" so it can travel faster. Along the way, the packets are separated. Every time a packet reaches a new hub, switch or router, it is sent on the fastest route available.

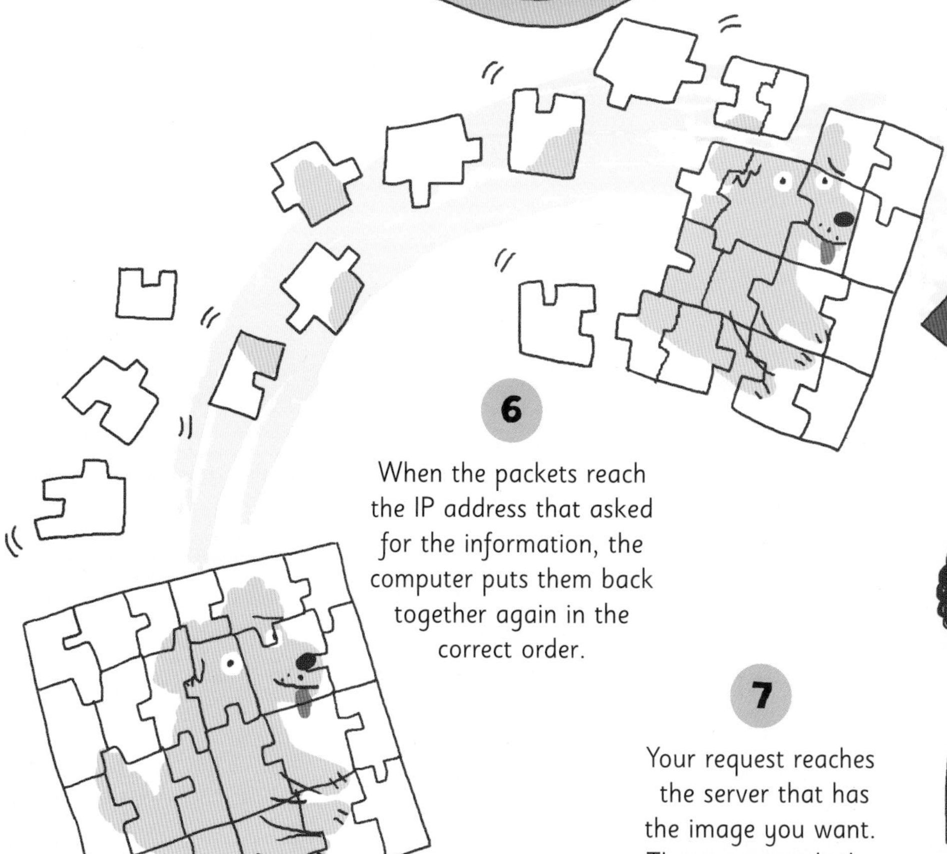

6

When the packets reach the IP address that asked for the information, the computer puts them back together again in the correct order.

7

Your request reaches the server that has the image you want. The server sends the image to your computer. This all happens in less than a second!

From studio to stereo

The journey of music

What's your favorite song? How do you listen to it? Does the singer perform it in your room? (It's unlikely although not impossible.) So how can you hear it as though the singer was in the room with you? How was that song captured?

Sound waves travel a certain distance before running out of energy. If the sound waves reach you, the sounds are "within earshot." If they cannot reach you, you're "out of earshot."

RAREFACTION
COMPRESSION
WAVELENGTH

1

As a singer's vocal cords open and close, they move back and forth, vibrating very quickly. When they close, they compress (squash) the molecules in the air together. When they open, they let the air "rarefact" (spread out again). These squashing and expanding movements pass on from air molecule to air molecule, rippling out. This is called a sound wave.

2

A microphone has a cone-shaped piece of plastic called a diaphragm in it that works a bit like vocal cords in reverse. The diaphragm vibrates with the rise and fall of the sound waves and converts those vibrations into electrical signals that rise and fall in the same pattern. The electrical signals can be recorded in different ways—**analog** or **digital**.

RECORDING

The best way to capture as many sound waves as possible is to get right in front of their source. Your singer was probably in a booth in a recording studio, a tiny room that traps the sound waves, with a microphone right in front of their mouth. The booth traps the sound waves and deflects them back again so that the microphone can capture as many as possible and get the full impact of the singer's voice.

38

3A
ANALOG RECORDING

1: Recording the whole sound wave is called analog recording. A vinyl record uses the electrical signals from the microphone to control a needle. The needle creates the jagged pattern of the sound waves in a spiral groove on a plastic disk—the record. The sounds are stored on the disk, exactly as they were made when they were produced.

2: Do you listen to songs on a record player? To play the sounds, the record player's needle moves along the groove. The wavy pattern in the groove makes the needle vibrate. A device called an amplifier turns the vibrations into electrical signals.

Records are an old-fashioned way of reproducing sound, but many people think they give the best recording of the original.

3B
DIGITAL RECORDING

1: You might be listening to your favorite song online or as a download from a computer. This is a digital recording. It was made by taking samples of the wavy pattern of the sound waves created by the singer—around 44,000 times per second—and mapping the highs and lows of the waves onto a graph. On the graph, each sample point has a coordinate—a number that a computer can read and understand.

2: When you tell your device to play that song, it takes the numbers on the graph and converts them into electrical signals.

4
The electrical signals are converted into sound waves by going through the diaphragms of the speakers, which vibrate in the same way as a diaphragm on a microphone. If you have more than one speaker (for example, earphones), the sound waves come at you from different directions, seeming to surround you—this is known as "stereo." Now you're listening to your favorite music as if the singer was in the room with you!

The higher the sample rate or "hertz"— the number of samples of the sound-wave patterns per second—the more accurate the digital recording will be.

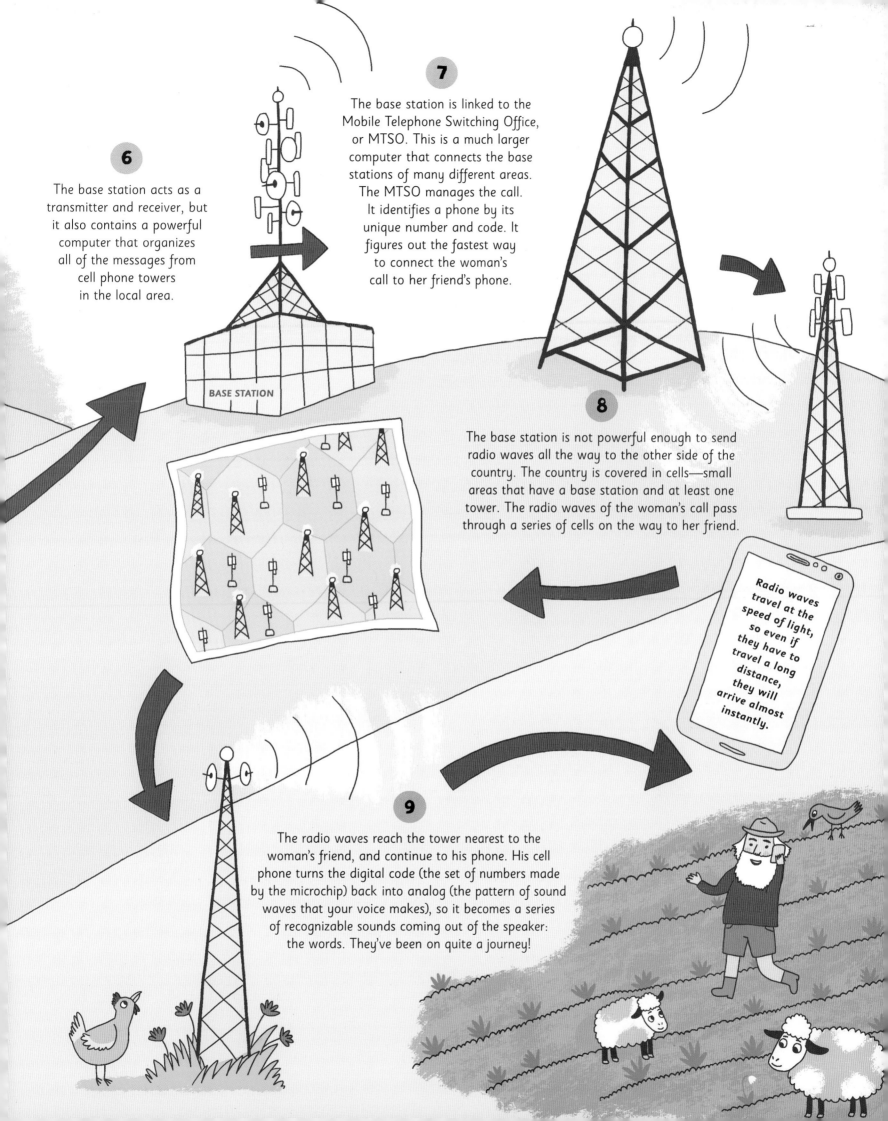

HOW DOES Electricity Reach My House?

The journey of electricity

We use electricity with barely a thought—flick a switch and it's there! But where does it come from? Electricity is a form of energy. In the early 19th century, British scientist Michael Faraday discovered he could create electricity. If he moved a magnet around a loop of wire, the wire became electrified. This is called "electromagnetic induction." We still produce electricity in the same way.

1

Electricity is produced by power plants all around the world. Inside are many enormous generators ("generate" means to create). They work in the same way as Faraday's experiment, except now they have giant magnets turning around massive wire loops.

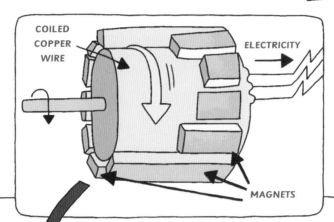

2

First, power is needed to move the turbines that turn the magnets. Turbines are machines which capture energy from moving water or steam. Different types of power plants use different primary sources of energy to create electricity.

A
FOSSIL-FUEL POWER PLANTS
burn coal, natural gas or oil from the ground to boil water. When water boils, it produces steam. The steam rotates the turbines of the generators. This is the cheapest form of fuel and produces most of the world's electricity.

B
WIND AND HYDROELECTRIC POWER PLANTS
use the power of moving wind and water to turn the turbines of the generators.

C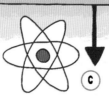
NUCLEAR POWER PLANTS
split atoms. This creates heat to boil water and produce steam to turn the turbines of the generators.

D
GEOTHERMAL POWER PLANTS
use heat from the center of Earth to boil water that produces steam to turn the turbines. This is only possible in places where this heat is near the surface of Earth.

E
SOLAR POWER PLANTS
convert energy from the sun's rays directly into electricity. But adapting the electricity grid to work with solar power plants is expensive, so there aren't many of them.

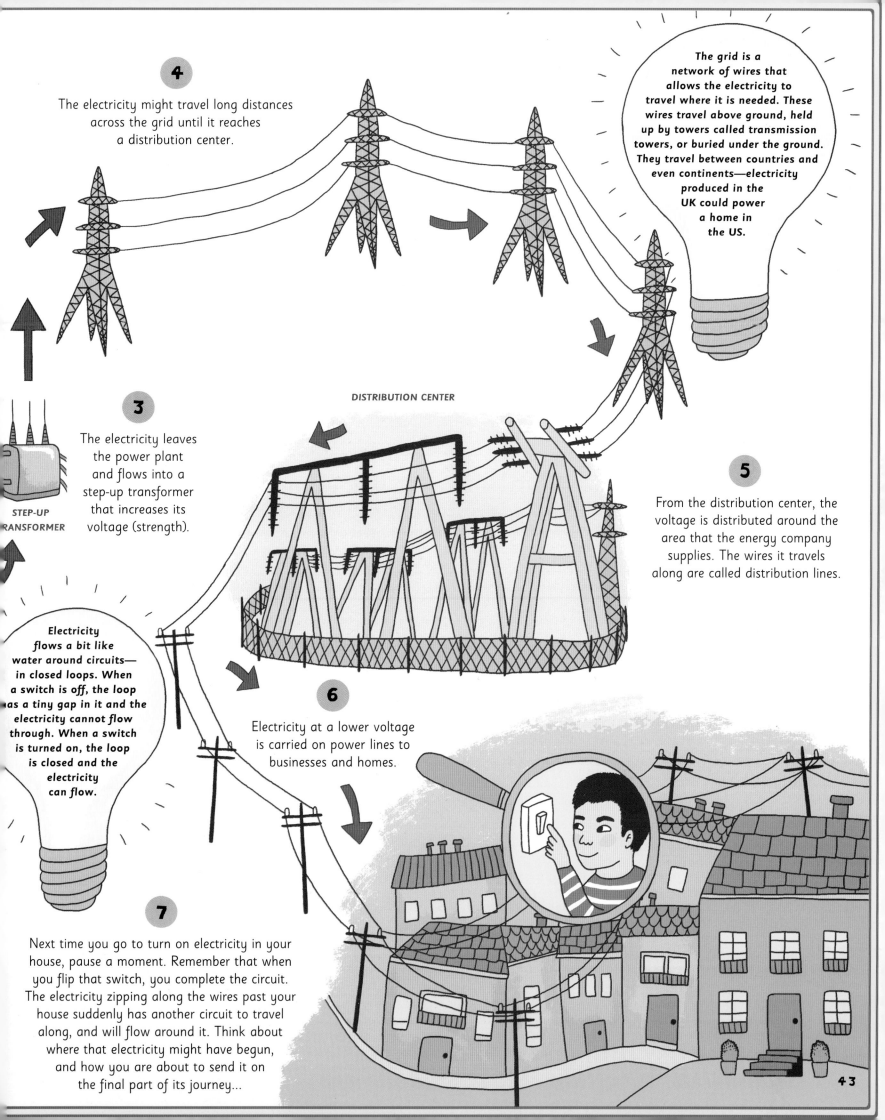

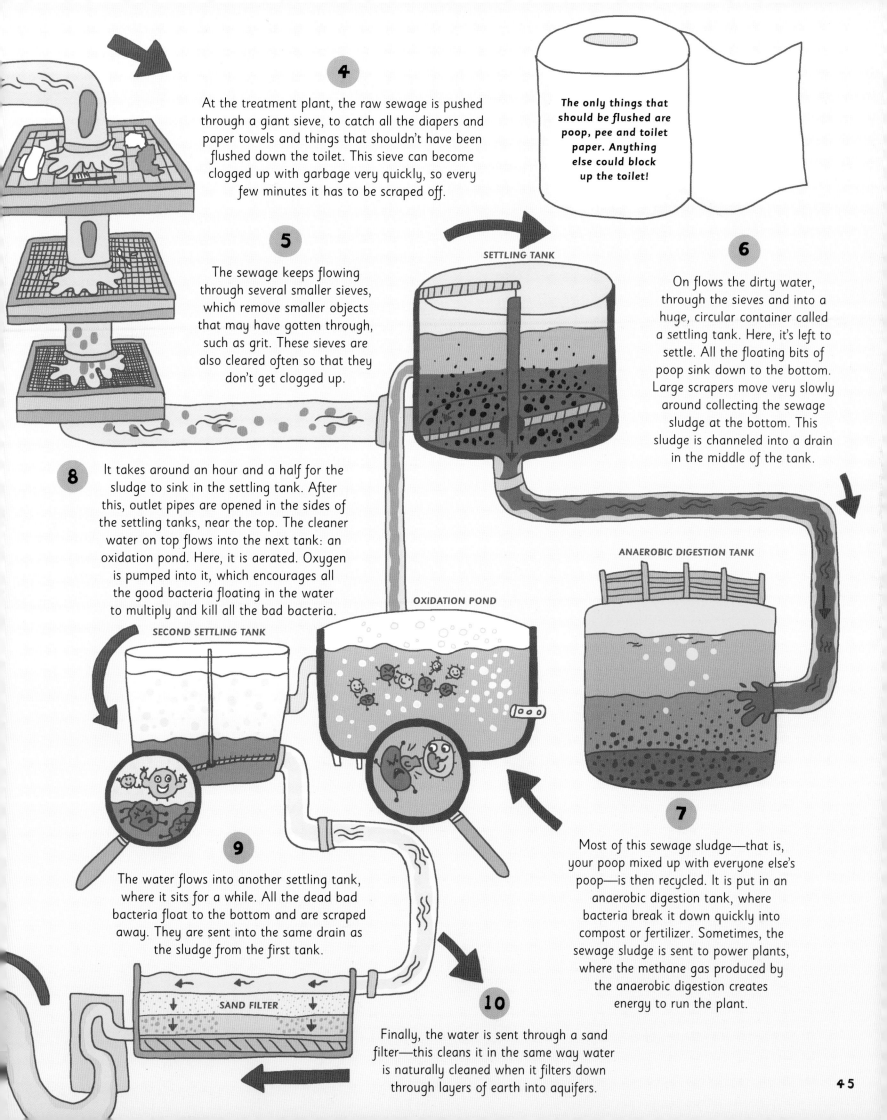

From cow to carton
The journey of milk

We all know that milk comes from a cow, and that it is used to make cheese, butter and other dairy foods. But when you pour that carton of milk over your cereal, do you ever think about the journey it has made to get there?

1

Milk is made in a cow's body from food and water. Dairy cows drink about a bathtub of water and eat about 100 pounds of food every day—the weight of 1,500 servings of your breakfast cereal. They mostly eat grass, hay and other crops with nutrients in them. The nutrients from the food go into their milk. It takes a cow about two days to turn food into milk.

Like humans, cows only produce milk after having a baby. Dairy cows have one calf every year to keep up their supply of milk, which lasts for around 10 months after their baby is born.

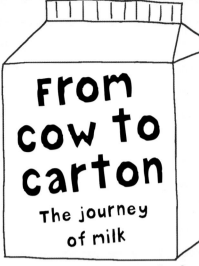

CONTINUOUS PASTEURIZATION SYSTEM

8 The raw milk is heated to kill any bacteria and then cooled. This is called pasteurization and makes the milk last longer.

First, the milk is pumped to a machine with a series of hot, thin metal plates. The milk flows along them for 30 minutes.

Next, the milk flows through a series of ice-cold metal plates, which instantly cool it.

Heating milk changes its taste. The temperature must be high enough to kill the bacteria, but must stay high for as little time as possible. Most dairies produce "high-temperature, short-time" (HTST) milk. It's heated to 161°F for 15 seconds. This produces the best taste and makes the milk last several days. Ultra-heat-treated (UHT), or long-life, milk has been heated to an "ultra-high temperature." It lasts for a long time, but tastes different than fresh milk.

9 The cooled, pasteurized milk is piped to the filling station, where a machine runs empty cartons or jugs along a conveyor belt, fills them with milk and stamps a heat-sealed lid on the top to prevent any bacteria from entering the carton.

10 The cartons are collected by large refrigerated trucks and delivered to stores nearby, ready to be bought and poured on breakfast cereal!

46

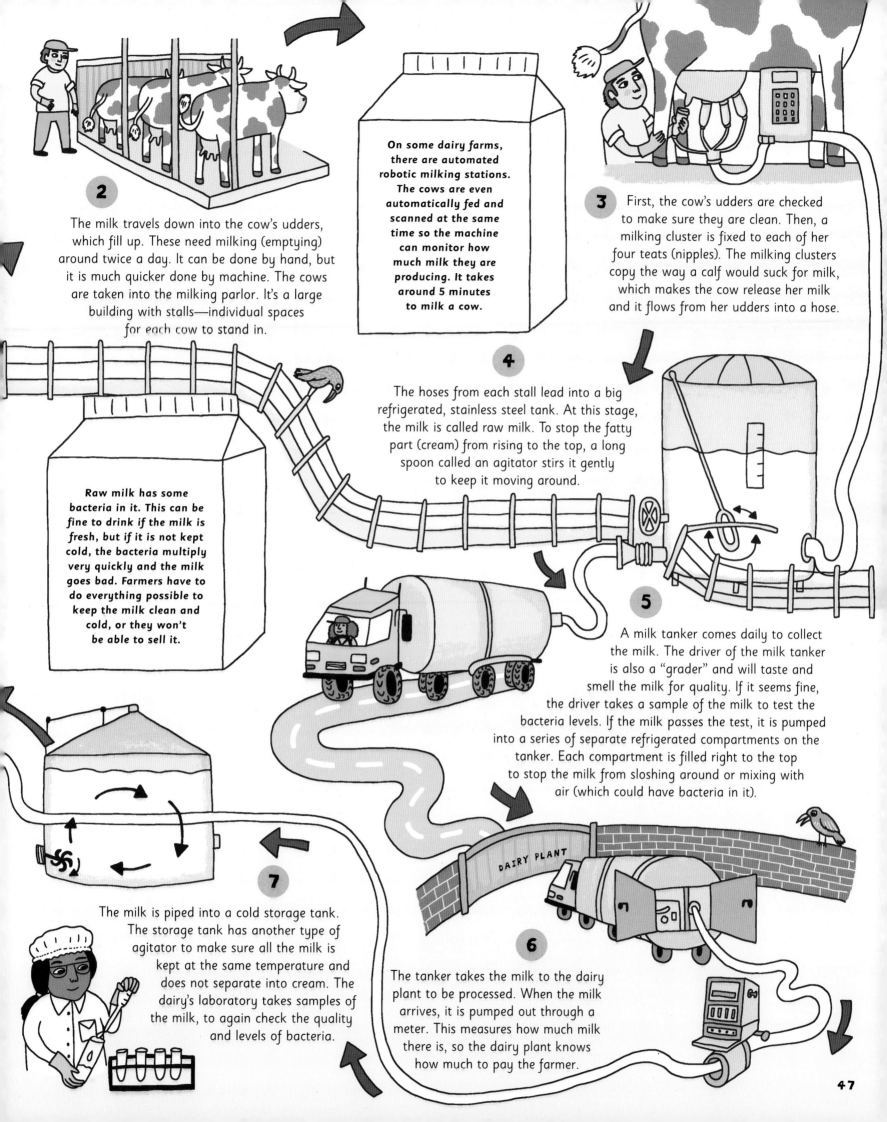

THE END

So, you've finished this book and come to the end of one journey.

Look around and try to notice the things you might take for granted. Think of all the machines, people, plants and processes that it takes just to make a chocolate bar.

Imagine what it may take to blast an astronaut into space!